Report of Investigations 9685

Man Mountain's Refuge: Refuge Chamber Training Instructor's Guide and Trainee's Problem Book

Michael J. Brnich, Jr., CMSP, Charles Vaught, Ph.D. (ret),
Kathleen M. Kowalski-Trakofler, Ph.D.

DEPARTMENT OF HEALTH AND HUMAN SERVICES
Centers for Disease Control and Prevention
National Institute for Occupational Safety and Health
Office of Mine Safety and Health Research
Pittsburgh, PA • Spokane, WA

July 2011

This document is in the public domain and may be freely copied or reprinted.

Disclaimer

Mention of any company or product does not constitute endorsement by the National Institute for Occupational Safety and Health (NIOSH). In addition, citations to Web sites external to NIOSH do not constitute NIOSH endorsement of the sponsoring organizations or their programs or products. Furthermore, NIOSH is not responsible for the content of these Web sites.

The findings and conclusions in this report are those of the author(s) and do not necessarily represent the views of the National Institute for Occupational Safety and Health

Ordering Information

To receive documents or other information about occupational safety and health topics, contact NIOSH at

> Telephone: **1–800–CDC–INFO** (1–800–232–4636)
> TTY: 1–888–232–6348
> e-mail: cdcinfo@cdc.gov
>
> or visit the NIOSH Web site at **www.cdc.gov/niosh**.

For a monthly update on news at NIOSH, subscribe to NIOSH *eNews* by visiting **www.cdc.gov/niosh/eNews**.

DHHS (NIOSH) Publication No. 2011–195

July 2011

SAFER • HEALTHIER • PEOPLE™

Contents

Instructor's Guide

Exercise Summary 1
How to Use This Exercise 2
Performance Objectives for Man Mountain's Refuge: Mine Refuge Chamber Training Exercise 3
    Objectives 3
    Intended Audiences 3
Instructor's Discussion Notes and Teaching Points 4
    Additional Points for Discussion 9
Answer Key for Man Mountain's Refuge Exercise 12
References 13

Trainee's Problem Book

    Instructions 14
    Training Exercise 15
    Problem 16
    Questions Pertaining to the Exercise 19
    Finding Your Score 32

Figures

Figure 1. 9 Right section map – face 17
Figure 2. 9 Right section map – outby 18

Acronyms and Abbreviations

CM	continuous miner
CO	carbon monoxide
NIOSH	National Institute for Occupational Safety and Health
SCSR	self-contained self-rescuer

Units of Measure

ft	foot or feet
lb	pound
ppm	parts per million

Man Mountain's Refuge: Refuge Chamber Training Instructor's Guide and Trainee's Problem Book

Michael J. Brnich, Jr., CMSP, Charles Vaught, Ph.D. (ret),
Kathleen M. Kowalski-Trakofler, Ph.D.

Office of Mine Safety and Health Research
National Institute for Occupational Safety and Health

Exercise Summary

This instructor's guide is designed for use by instructors who train mine employees on how and when to use a mine refuge chamber, and aids the instructor in reinforcing the critical decisions that have to be made during a mining emergency. The discussion notes and teaching points included in this instructor's guide are based on a paper-and-pencil simulation exercise that trainees use to learn about the choices that must be made in an emergency situation. In this exercise, trainees work through an interactive story that presents a scenario in which a section crew, along with additional general labor workers, must decide what to do when they learn there is a fire somewhere in the mine, but do not know the exact location. One of the characters in this story is Man Mountain, a member of the labor crew. As time goes by, the miners face a series of choices about how best to increase their chances for survival. The story is taken in part from real-life incidents. The teaching instructions in this instructor's guide have been designed for use with the simulation exercise, which is included in the Trainee's Problem Book.The completion of this exercise can help new miners, experienced miners, trainers, and others, who must deal with issues of self-rescue and escape, to become more aware of: (1) the need to gather as much information as possible as early as possible; (2) the value of knowing one's escapeways; (3) the need to use self-contained self-rescuers (SCSRs) properly; (4) the value of a multigas detector in an emergency; (5) when, and under what circumstances, to enter a refuge chamber; and (6) how to recognize the reaction signs of traumatic incident stress.

How to Use This Exercise

The following guidelines will help you, as the instructor, direct the training session:

- Inform trainees that the mine in the exercise may be laid out different from their own mine. Discuss the mine layout with the class and highlight specific differences.

- Define any terms with which trainees may not be familiar.

- Have trainees read through the Trainee's Problem Book and, for each statement, select "YES" for agree or "NO" for disagree.

- After trainees have completed the exercise, have them look at the answer key (which is provided in this guide). **Be sure trainees do not look at the answer key before they complete the exercise.**

- The exercise should take about 30 minutes to administer and 30 minutes for discussion. Use the training materials in one of the following ways:

  o During annual refresher training to introduce and discuss the decisionmaking process associated with the use of portable and permanent refuge alternatives.

  o As part of mandated quarterly escape drills.

- You will need the following materials to administer the exercise:

  o "Man Mountain's Refuge: Trainee's Problem Book" for each trainee (which can be copied from this guide).

  o An exercise answer key for each trainee (which can be copied from this guide).

  o This document, "Man Mountain's Refuge Instructor's Guide and Trainee's Problem Book," with answer key and discussion notes for each instructor.

Performance Objectives for Man Mountain's Refuge: Mine Refuge Chamber Training Exercise

Objectives

After the completion of the exercise, new miners, experienced miners, trainers, and others who must deal with issues of self-rescue and escape will be able to:

- Explain the need to gather as much information as possible as early as possible
- Explain the value of knowing one's escapeways
- Understand the need to use self-contained self-rescuers (SCSRs) properly
- Discuss the value of a multigas detector in an emergency
- Identify when, and under what circumstances, to enter a refuge alternative
- List several normal psychological and physical responses of individuals in an emergency

Intended Audiences

- Underground coal miners in annual refresher classes
- Contractor employees
- Trainers
- Responsible persons at mines
- Command center personnel including mine managers
- Manufacturers of mining equipment
- Researchers

Instructor's Discussion Notes and Teaching Points

Use the information presented here and on the answer key (in this guide), your own ideas and experience, and those of the trainees in your class to discuss the exercise after it is completed. Group discussion can strengthen knowledge and skills, correct errors, and relate the exercise content to the experiences of the miners. After they have worked the exercise, miners often enjoy discussing the problems from the exercise. They also frequently think of better ways to respond to a problem than those listed among the answers. The purpose of the exercise is to help miners think about and remember basic knowledge and skills they may someday need to deal with in an emergency. This type of discussion can reinforce those concepts and tailor the exercise content to the needs of the group you are training.

The following notes provide the correct answer(s) for each question as well as additional information for you to discuss with your class. Read through and think about these notes before the class. Do not read the notes to the class members. This would be boring and ineffective. Rather, incorporate the ideas you find here with your own ideas and make these points at the appropriate times during the discussion of the exercise. The numbers in parentheses ( ) indicate the statement answer numbers that are found below each question.

Question A. The correct statement is **3**. One miner should warn the face crew while another notifies the mechanics and bull gang members on the section. Then everyone should meet at the dinner hole. There are at least 18 miners on the section. They all need to be informed about the situation, and everyone needs to meet at the dinner hole to make sure they are all accounted for.

Notifying the face crew while one miner looks for the foreman (1) is not enough because all miners must be told. Going to the dinner hole immediately (2) is not an option because the other miners have to be notified. Although the shuttle car operator may be a joker, ignoring his message (4) is the wrong thing to do as there could be a serious problem. It is not necessary to don SCSRs and start out the return (5). In responding to question A, several points must be stressed by the instructor. First, the other miners must be notified of the situation. Second, there is no smoke in the section at this time. Third, if there is an emergency, it is dangerous for one miner to proceed alone out the alternate escapeway.

Question B. The correct statements are **8, 9, 10, 11,** and **13**. Extra SCSRs should be distributed immediately to each miner (8) and someone should retrieve the tether rope that can be used as a lifeline (9) between miners evacuating on foot and encountering heavy smoke. Power to the section needs to be cut (10) and a head count needs to be taken to be sure all miners are accounted for (11) once everyone is at the dinner hole. Someone should also start the mantrips and bring the outby mantrip closer to the dinner hole (13).

There is no need to send everyone to the pager phone to find out what's going on (6) because the section foreman is at the phone attempting to find out what is happening. Besides, he could be easily distracted and miss hearing important information if all the miners show up at the phone. While it is tempting to head for the mantrips immediately (7), miners should make preparations before leaving the section. Similarly, some miners may become excited and want to get the section foreman immediately and evacuate (12). Prior to leaving, the section foreman must make his best attempt to find out the location and nature of the fire.

Question C. The correct answers are **14** and **16**. Miners need to don their SCSRs now before getting into the mantrips (14) because someone has smelled smoke. Once the crew starts out of the section, the mantrips would have to stop and everyone would have to get out to don their SCSRs. It is also important to inform the section foreman that there is smoke coming into the section (16). Because he is at the phone, he may not have smelled the smoke yet.

It is not a good idea to have one of the mantrips start off the section (15). Doing nothing while waiting for the section foreman to come back with more information is unwise (17). Someone needs to try to find out the location of the fire. The crew has already spent several minutes waiting to see what is going on. Anything that can be done to prepare to escape should be done.

Question D. The correct answer is **19**. Because wispy smoke is coming onto the section and there is still no word on the nature or location of the fire, you need to tell the section foreman it's time to start out of the section before the smoke gets heavier (19).

Staying at the phone to wait for more information while the foreman goes back to the mantrips to stay with the crew (18) wastes valuable time. While checking the returns and the secondary escapeway for smoke (20) might not be a bad idea, it is a poor use of time. Even if these entries were clear, the crew would have to walk more than a mile to reach the permanent refuge chamber, let alone the mains. While it would be helpful to have more information about the fire, it is pointless to return to the mantrips and wait (21). Smoke is already on the section—it is time to leave.

Question E. The correct answer is **22**. Because the smoke was heavier at the permanent refuge chamber and there was still no new information on the fire, the best option is to have the crews get into the permanent refuge chamber and wait for the mine rescue team (22).

Having the crews move into the secondary escapeway (23) or return (25) to continue escape on foot is not a good idea because miners would have to travel almost a mile uphill just to reach the mains. Staying on the mantrips and continuing out to the main permanent refuge chamber in North Mains or to the outside (24) is also a possibility. But because the smoke is getting heavier and there is still a considerable distance to travel, it is best to go to the safety of the permanent refuge chamber.

Question F. The correct answers are **26, 27, 29**, and **30**. Once the mantrips cannot go any farther in the thick smoke and miners begin leaving the mantrips, it is important to get the miners back together (26). Because it is wise to go back to the permanent refuge chamber, someone needs to check the alternate escapeway for smoke (27). Chances are the smoke is lighter in this entry than in the primary escapeway and it would be easier to travel in the entry. All the extra SCSRs (29) should be taken along. The tether rope should be used as a link line between miners (30). This will help keep everyone together while heading inby to the permanent refuge chamber.

Bringing up the rear while the boss leads the crew out the primary escapeway (28) is not a good option. The crew would have to travel about a mile uphill in heavy smoke just to reach the mains.

Question G. The correct answer is **31**. At this point, a difficult choice has to be made. The section foreman and several other crew members want to continue escaping by following the secondary escapeway out while one miner says he cannot continue escaping uphill on foot. The best choice is to have everyone go back to the permanent refuge chamber and wait for the mine rescue team. However, each miner can decide whether or not they want to go to the chamber, so it is important to ask who is going to the chamber (31). It is important to note that you cannot mandate the crew to do what you think is best. Each miner is in a situation of choice, but splitting up the crew is often not the best choice. In this case, going to the refuge chamber is the safest thing to do because there is fresh air and communications with the outside.

Taking a vote on whether to go back to the refuge chamber or continue is a poor choice (32). This is a waste of time. Staying back to help Sam, the mechanic, while the crew proceeds out on foot is not an option (33). Each miner must act based on their best judgment and, hopefully, the understanding that going to the permanent refuge chamber is the best course of action. Moving the road grader so the mantrip can continue (34) is also not a good choice because the smoke is so dense making it difficult to continue travel in the intake.

Question H. The correct answer is **37**. Going to the permanent refuge chamber by grabbing the lifeline and using the secondary escapeway is the safest action in this situation.

Getting back on the mantrip and riding back to the chamber (35) is not practical in the dense smoke. Walking back to the chamber through the primary escapeway (36) is not an option because the smoke is very dense in the intake and it is lighter in the secondary escapeway. Deciding to follow the section foreman out (38) is a wrong choice because the one miner has already decided that taking refuge in the permanent refuge chamber is the best option.

Question I. The correct answers are **39, 40, 41,** and **43**. Once all the miners are inside the refuge chamber, there are several things that must be done. First, call outside to check on the fire, report the names of everyone in the chamber, and advise the communications person that the foreman and others are coming out on foot (39). You need to know the status of the fire, and personnel outside need to know the location of the crew. Miners can take off their SCSRs (40) because they are in the safety of the permanent refuge chamber which has positive pressure intake air from the surface. The maintenance foreman should occasionally check the air in the chamber for oxygen deficiency, methane, and carbon monoxide using a multigas detector (41). Although there is fresh air into the chamber, the atmosphere needs to be checked in case of any leakage. Miners should be reminded to turn off their cap lamps to conserve battery power (43) because they may be needed later. One cap lamp should be enough for the time being.

While miners are probably hungry and thirsty, they should not open provisions and help themselves (42). Food and water must be conserved because it is unknown how long they will be in the chamber.

Question J. The correct answers are **46** and **47**. Man Mountain needs to stay with everyone in the safety of the chamber. He needs to be reminded that he would have to travel a considerable distance uphill through smoke just to reach the mains (46). Also, Man Mountain and the rest of the miners should be reminded that the section foreman and the other crew members who decided to travel on foot have not been heard from and that leaving now would be taking a big risk (47).

Checking the primary and secondary escapeways for smoke and leading the remaining miners out (44) would be dangerous because it is still about a mile uphill. If Man Mountain is determined to leave, the rest of the miners need to stay in the chamber (45).

Question K. The correct answers are **49**, **51**, and **52**. It is important to normalize the feelings and behaviors of the miners (49). Noting the obvious, that this is a stressful situation for everyone, helps put the potential anxious reactions individuals experience into perspective. Mentioning some of the normal symptoms that people have in such situations, both emotional and physical, can be reassuring. (51). Suggesting everyone take three deep breaths provides a physical/behavioral intervention that can divert attention and reduce anxiety, and also increase oxygen in their bodies (52). Other suggested interventions might be tensing and relaxing different parts of the body (arms, shoulders, etc.) or doing neck rolls.

Ignoring these behaviors and feelings will not help individuals calm down (48). In fact, ignoring them can contribute to their increase. Previous research on human response in emergencies has shown that most people do NOT tend to panic in emergency escape. They later report feelings of anxiety, but their behavior tends to be rational. Performing CPR on Shorty is not needed because he is breathing and talking (50).

From a psychological perspective, preparation is the key to mitigating negative responses. Knowing how normal people react in a stressful situation is an important part of training for mine emergencies.

Question L. While the answers to this question are open-ended, the following points should be discussed with the trainees:

> **Answer 53.** Why should individuals seek refuge in an emergency? The decision to seek refuge or not is going to be specific to the given incident. In the scenario presented here: (1) you do not know the location of the fire; (2) there is heavy smoke coming in that will inhibit your ability to successfully escape; and (3) you have a competent mine rescue team member with the crew. Therefore, the decision to go into the chamber is a logical choice.

> **Answer 54.** Distance needs to be considered when making the decision to seek refuge. Trainees may need to be reminded that the miners had about 5,000 ft to travel uphill through dense smoke just to reach the mains. During this time, they would have had to rely on their SCSRs. Although a cache of SCSRs was located at the permanent refuge chamber allowing miners to switch to fresh units, the next cache would be located at the mouth of the section. Even if the miners are well trained in using their SCSRs, walking uphill wearing an apparatus (SCSR) would probably prove difficult for some.

Answer 55. Going to the permanent refuge chamber will get the miners out of the smoke. It will also get them to a location where there is (1) fresh water and food; (2) communication with the outside; and (3) a source of fresh air.

Answer 56. Possible reasons for not seeking refuge: (1) Since the advent of SCSRs, miners have been taught to escape. Even with the availability of refuge alternatives, including permanent refuge chambers, the miners' first line of action is to attempt escape. Therefore, the common thought today is to escape rather than seek refuge. (2) By choosing to go to a refuge alternative, miners can run the risk of conditions going from bad to worse. (3) There are psychological considerations as well. Knowing the location of the fire or encountering only light smoke would be reasons why one would not seek refuge. However, the decision to seek refuge or not will depend on the situation.

Question M. The correct answers are **58**, **59**, and **61**. After a trauma, individuals frequently report having difficulty concentrating and may experience interrupted sleep. This can be a safety issue particularly when driving (58). It is not a good idea to make major life decisions soon after a trauma (59). Giving yourself time and space will be helpful to feeling better. This may include requesting light duty or outside work for awhile. Taking extra care of oneself through adequate sleep, good nutrition, and physical activity is important. It does get better. If an individual continues to feel anxious or depressed and day-to-day functioning is still a problem three months after the incident, it is time to seek a *disaster* mental health counselor (61).

Additional Points for Discussion

Between March 1988 and November 1990, three major fires occurred at underground coal mines in the eastern United States. These fires resulted in the permanent closure of two of these mines and forced a total of more than 60 miners, caught inby the fires, to evacuate considerable distances through moderate to heavy smoke. NIOSH researchers had the opportunity to interview 48 of the miners who escaped these fires and talk with them about their experiences. Information provided by these individuals revealed that escaping the mine fires posed problems for many of them.

Throughout this exercise, finding out the location and nature of the fire is a critical point that must be emphasized with the trainees. The reason is that few of the 48 fire escapees knew the location of the fire at their mine. Although someone at each of the three mines knew almost immediately where the fire was, only 2 miners took the time to find out the location of the fire.

Lacking this critical piece of information presented problems for many escapees. For example, many miners delayed donning their SCSRs and traveled through smoke without this protection because they did not know how far they had to travel and wanted to save their SCSRs in case they were needed later. As a result, some two thirds of the miners interviewed said they donned their SCSR in smoke, and many reported difficulty donning their units in the smoke. Others reported difficulty breathing with the SCSR because they had been traveling on foot before putting the device on and were breathing heavily when they donned the unit. One miner said:

> ... [the] man that's in charge needs to take his time and walk out of there slowly and easy with his self-rescuer on ... If you go 6, 7, 800 feet before you even try to put one of them things on, you're winded. [Then], it's like trying to suck through a straw.

In hindsight, a number of miners said that, in the future, they would don their apparatus before boarding the mantrip and leaving the section. Discuss with the class the importance of donning SCSRs before boarding the mantrip.

Similarly, lack of information on the location of a fire influenced miners' escape strategies. One miner said of his crew when they were discussing strategy:

> We were going to try ... getting to Peterson [air shaft], but we didn't know exactly where the fire was. We thought that the fire was at 3 Left, Number 2 [belt] transfer ... So that was our idea ... I wished we knew where the fire was for one thing. It's like you're going into the unknown; you don't know exactly where you're going.

At least four groups of escapees attempted to ride out of the mine on mantrips or other vehicles. One crew traveled about half a mile before the smoke became too thick and the crew had to continue traveling on foot. After leaving their mantrips, though, nearly all miners encountered some level of difficulty once they entered their escapeways. Discuss with miners the pros and cons of traveling as far as possible on the mantrip before having to stop due to heavy smoke or other problems.

Miners must be aware of the fact that smoke was the greatest physical problem with which escapees had to cope. Escapees said that visibility ranged from less than 2 ft in the primary escapeways and in the track and belt entries to as much as 60 ft in the return air courses (some of which were designated as secondary escapeways). Some miners did not expect the smoke that they encountered to be as thick as it was:

> I didn't expect it to be that thick ... they show you movies that you can get down ... and crawl out, but I don't think you could do that ...

Miners characterized the smoke in different ways by color and thickness. In an area where the smoke was lighter, a worker described it as having a bluish-gray color and was "like ... more just like a filtering smoke." Other miners encountered thick, heavy smoke. A utility man said "You couldn't see ... it was just like ... like riding in behind a bulk duster for rock dusting." Another miner said:

> The stopping was probably on the other side of the props, but I couldn't see it. I couldn't even see the door, that's how thick it was. I put my hands out ... and couldn't see the end of my fingers.

In health and safety training, miners are taught to follow the primary or secondary escapeways out of the mine to safety. For ease of navigation, mines mark these entries using different colored reflectors or signs. Escapees said it was hard if not impossible to see the escapeway markings because of the thick smoke. Therefore, it was easy for miners to become disoriented and lost during their escape. A former mine rescue team member recounted the effect the thick smoke had on him even with his experience:

> ... I thought we were walking right into the fire ... I started to get a little upset, a little tight ... And in our returns [secondary escapeways] we have reflectors ... And it's a good idea if there's no smoke ... [but if] you fall down and you get up and you get turned around, you know, if somebody doesn't know where you're going, you could be crawling around down there.

Some miners became confused and disoriented in the smoke during their escape. Although he followed the belt line from the section, one miner recounts his experience as he made his way from the belt to the track:

> ... as soon as I found the crosscut, I went in because I didn't want to miss it and I went to the end of the crosscut and run into a permanent stopping. Well, I started looking for the (man)door and it seemed like I was lost. I wasn't lost but it seemed like I was lost because I got sort of that feeling, well, I know that door is here but I just couldn't find it...

A mechanic, who followed the same belt line said:

> I didn't [seem to] know my way out of there. I lost all orientation [on] how to get out of there. I knew my way out, but I forgot.

One crew of miners became so disoriented in the smoke during their escape that they actually started back into the section. The section foreman recalled:

> ... they said they couldn't make it over those overcasts; there was too much smoke. So, we started back because I noticed a 3 x 3 [man]door in the return. So, I wanted to get back into the intake. Well, I couldn't find that 3 x 3 [man]door and I knew I didn't want to start running around in circles. ... we started up over an overcast in the return and in the ... sidewall ... there was a 3 x 3 door ... everybody went [through the door] out into the intake escapeway ... We started walking ... but something didn't look right to me ... [A]round vacation they had dug a sump and you had a path; as you come out your intake escapeway, the slate's on your left side and the path's on your right ... I'm walking along and I started thinking something's wrong because that ... slate should be on my left, not on my right.

Realizing the crew was headed back to the section, the section boss decided to go back into the return. He then led the crew to another overcast, opened a door, felt air on his face, and was able to determine which way the group should go from there.

Finally, at least one miner interviewed became physically exhausted during his escape. The miner, a CM operator who was over 6 feet tall and weighed more than 250 lb, was traveling with his crew as they followed the belt line off their section. The entry was 44 to 48 in high with a clearance of some 3 ft between the belt and rib. The crew traveled some 600 ft before encountering smoke and donning their SCSRs. They then traveled several hundred ft more when the miner began having trouble breathing. The foreman split the crew and allowed faster miners to continue while he and several other miners stayed with their buddy. In the end, the foreman and the others had to leave the CM operator behind and continued on outby the fire. The CM operator was later rescued and survived the ordeal. Have trainees discuss their thoughts about leaving a disabled miner behind.

While there was the potential for disaster with each one of the three mine fires, all miners escaped successfully and no lives were lost. At best, miners' escapes from the fires were difficult with some crews having to travel nearly two miles on foot to reach safety outby the fire areas. In addition, none of the eight crews had to travel up steep inclines during their escapes as the coal seams at all three mines were fairly level.

Since the enactment of the 2006 MINER Act, mines are required to have lifelines with directional cones installed in both primary and secondary escapeways. These lifelines are designed for miners to grab onto and use as a guide to find their way out. But the lifeline may be broken, or escaping miners may decide to follow another route other than a designated escapeway; discuss with trainees how they might deal with these situations when navigating in heavy smoke.

Answer Key for Man Mountain's Refuge Exercise[1]

The correct answers are marked with an asterisk.

Question	Answer Number						
A	1	2	3*	4	5		
B	6	7	8*	9*	10*	11*	12	13*
C	14*	15	16*	17			
D	18	19*	20	21			
E	22*	23	24	25			
F	26*	27*	28	29*	30*		
G	31*	32	33	34			
H	35	36	37*	38			
I	39*	40*	41*	42	43*		
J	44	45	46*	47*			
K	48	49*	50	51*	52*		
L	Open ended questions – not scored						
M	57	58*	59*	60	61*		

---

[1] This page may be duplicated and distributed to trainees for reference.

References

Brnich MJ, Vaught C, Mallett L [1992]. SCSR proficiency requires hands-on practice. COAL 97(7):52–54.

Mine Improvement and New Emergency Response Act of 2006 (MINER Act), Pub. L. No. 108-236 (S 2803) (June 15, 2006).

NIOSH [2007]. Research report on refuge alternatives for underground coal mines. U.S. Department of Health and Human Services, Centers for Disease Control and Prevention, National Institute for Occupational Safety and Health, Office of Mine Safety and Health. 16 pp. Available at: [http://www.cdc.gov/niosh/mining/mineract/pdfs/Report_on_Refuge_Alternatives_Research_12-07.pdf]. Date accessed: November 2010.

NIOSH [2000]. Behavioral and organizational dimensions of underground mine fires. By Vaught C, Brnich MJ, Mallett LG, Cole HP, Wiehagen WJ, Conti RS, Kowalski KM, Litton CD. Pittsburgh, PA: U.S. Department of Health and Human Services, Centers for Disease Control and Prevention, National Institute for Occupational Safety and Health, DHHS (NIOSH) Publication No. 2000–126, IC 9450.

NIOSH [2009]. Harry's hard choices: mine refuge chamber training, instructors guide. By Vaught C, Hall E, Klein K. Pittsburgh, PA: U.S. Department of Health and Human Services, Centers for Disease Control and Prevention, National Institute for Occupational Safety and Health, DHHS (NIOSH) Publication No. 2009–122, IC 9511.

71 Fed. Reg. 71429 [2006]. Mine Safety and Health Administration, 30 CFR parts 3, 48, 50, 75: emergency mine evacuation; final rule.

Trainee's Problem Book

(with instructions for exercise completion)

This is a story about an underground continuous miner (CM) section crew and additional general laborers who must decide what to do when they learn there is a fire somewhere in the mine but do not know the exact location. As time goes by, the miners face a series of choices about how best to increase their chances for survival. This story is taken in part from real-life incidents.

Instructions

Form a group with several of your coworkers. Then, read the story and the questions about the story that follow. Each question is followed by a series of statements relating to the question. Read each statement and decide whether or not you agree with the statement; then, discuss the statement with the group, decide on a group answer, and circle either **YES** or **NO** for each statement.

Once you have finished the story, look at a copy of the answer key, provided by your instructor. Compare your answers to the answer key, but do not change any of your answers. Discuss the story and answers with the class and your instructor.

Training Exercise

The Mining Story (Scenario)

You are a mechanic on the 9 Right CM section at Kathleen Mine. You are a member of the mine rescue team.

On this day, there are a total of 19 miners, including yourself, working on your section. These include the 11 regular production crew members, 4 mechanics and 1 maintenance foreman, plus 3 labor miners, including "Man Mountain" Jenkins, from the general labor crew. The rest of the labor crew is outby doing roadwork.

This is a five-entry section that has been driven 10,000 ft from the mains.

The coal seam is pitching and roughly 84 in thick on this section.

The 72-in belt is located in the No. 3 entry. The belt is ventilated by a separate split of intake air that travels inby and is dumped into the return through a pipe regulator just outby the tailpiece.

Everyone is wearing one-hour SCSRs on their belts. There are 12 SCSRs and a tether rope stored on the section. Two mantrips on the section each have 12 SCSRs and additional units are stored outby in caches located 30 minutes apart (or about 5,000 ft) between the primary and secondary escapeways.

The primary escapeway is in the No. 4 intake entry; the secondary escapeway is in the No. 3 belt entry. Both escapeways are equipped with lifelines.

A portable refuge chamber is located on the section at the No. 96 crosscut, outby the faces between entries No. 4 and No. 5 (see Figure 1). A permanent refuge chamber is located along 9 Right at the No. 50 crosscut between the primary and secondary escapeways, entries No. 3 and No. 4, (see Figure 2). Another permanent refuge chamber is located in North Mains between 9 Right and 10 Right.

The permanent refuge chambers have a borehole to the surface with downhole ventilation, pager phone, telephone communication to the outside, food, water, first-aid supplies, and toilet facilities.

Both the maintenance foreman and section foreman have multigas detectors. The mine uses rubber-tired, mobile diesel equipment for moving supplies and personnel.

The mine has wireless communication. The section and maintenance foremen both have portable two-way radios. Mine pager phones are located at the belt tailpiece, at each belt transfer, and at each portable refuge chamber or permanent refuge chamber.

Mandoors are located at approximately every five crosscuts.

Pillars are on 100-ft centers.

Now turn to the problem statement on the next page.

Problem

You are in the No. 4 entry working on the bolter. The shuttle car operator, who's been joking with you all shift, comes up from the feeder and tells you there's a fire on a piece of equipment somewhere in the North Mains. He says, "I'm going back to the phone!"

Look at Figures 1 and 2.

Then go to **Question A** and begin.

Work on one question at a time.

Do not jump ahead to any other questions, but you can look back at previous questions anytime.

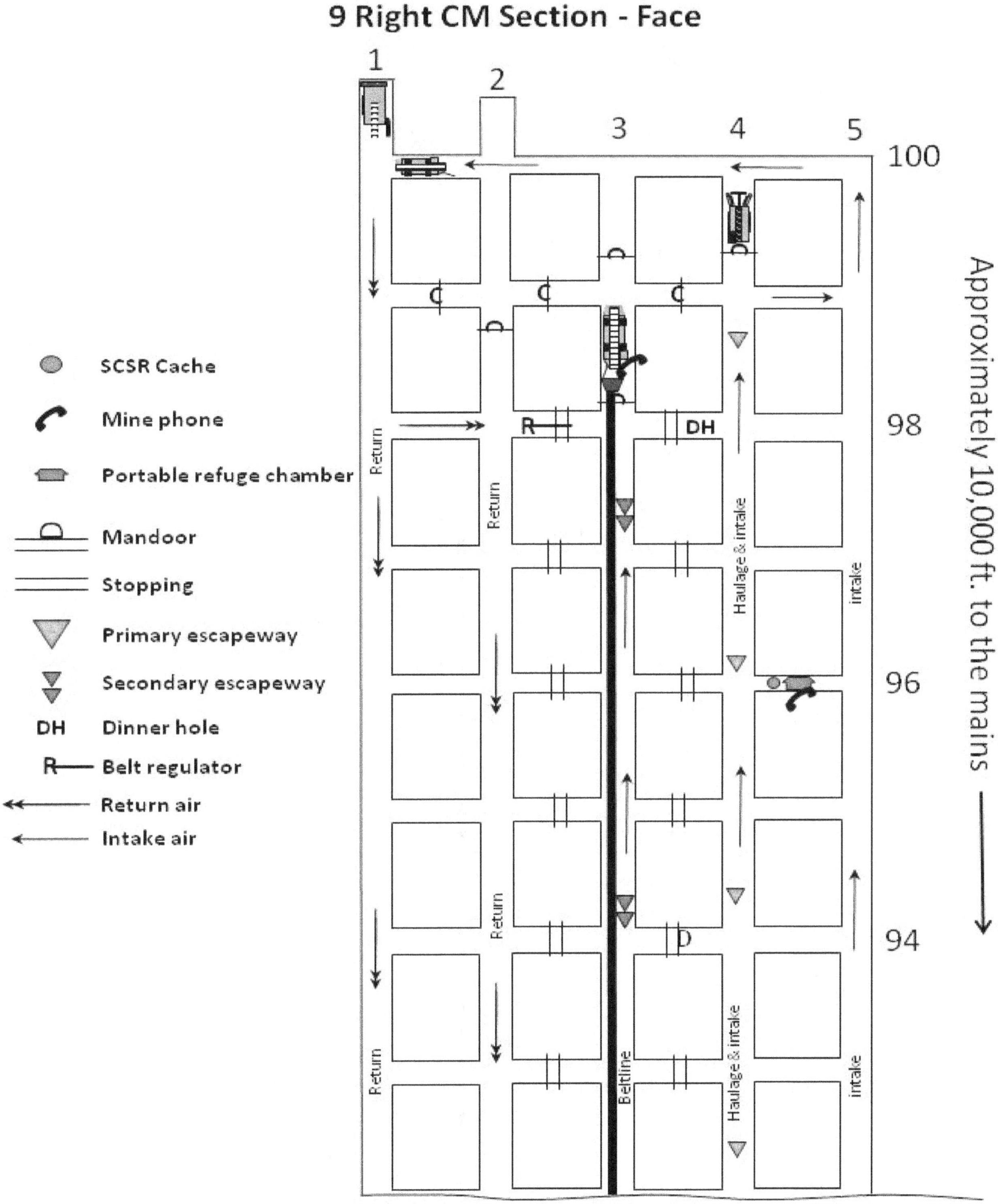

Figure 1. 9 Right section map – face

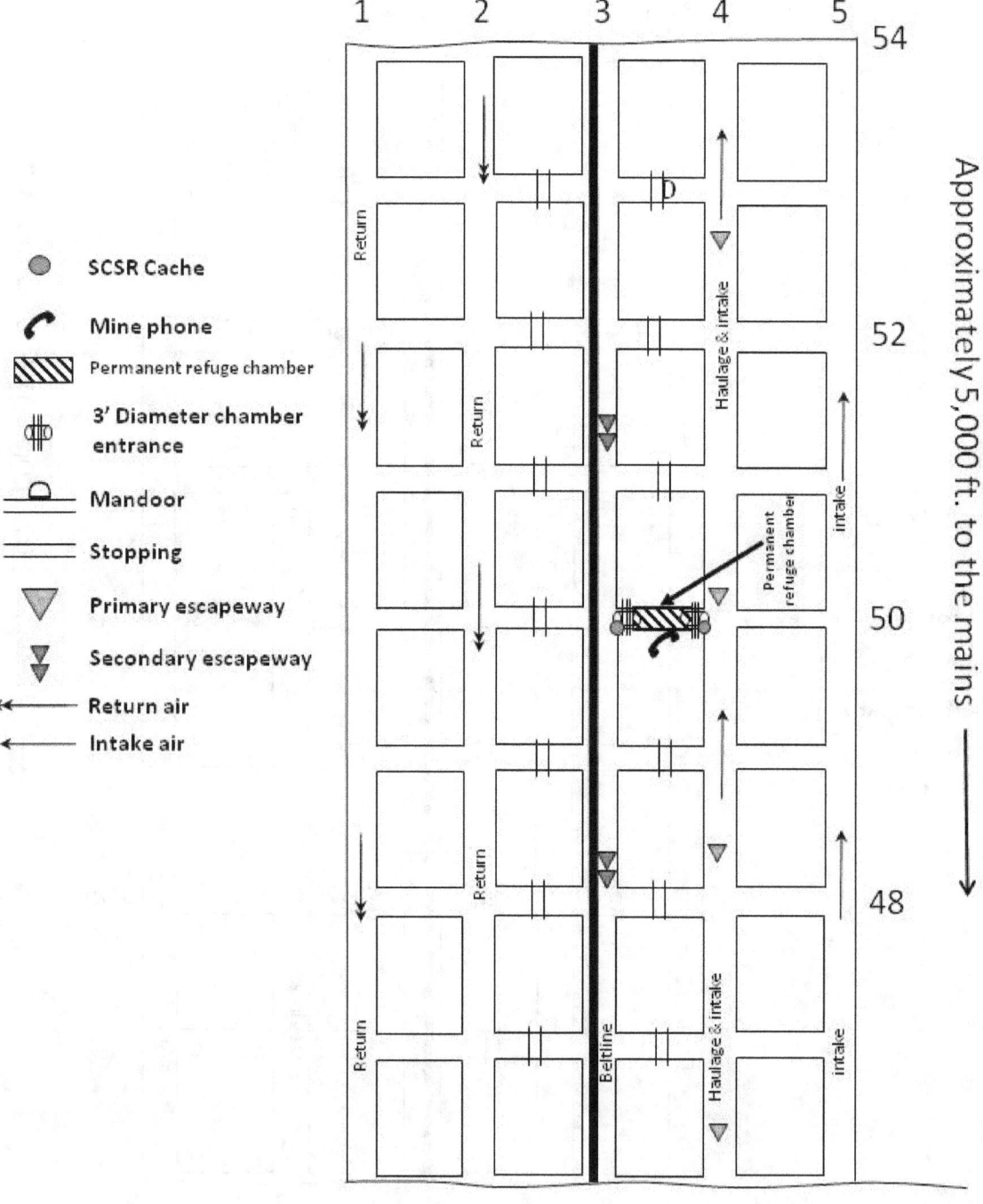

Figure 2. 9 Right section map – outby

Questions Pertaining to the Exercise

Question A

The shuttle car operator heads back to the tailpiece. Your buddy, who has been helping you on the bolter, asks, "What do you think we oughta do?" You reply, "I think we'd better start moving!" What should you do?

1. YES NO   Ask your buddy to notify the face crew to gather at the dinner hole while you go look for the section foreman.

2. YES NO   Tell your buddy that you should go immediately to the dinner hole.

3. YES NO   Ask your buddy to notify the face crew while you get the other mechanics and the labor crew workers. Tell him you'll meet in the dinner hole.

4. YES NO   Ignore the shuttle car operator and finish up what you're doing.

5. YES NO   Put on your SCSR and proceed directly out the return.

When you have responded to all the statements above, go to the next page.

Question B

Your buddy has gone to the face and alerted the CM crew. You have notified the other mechanics and gathered the labor crew workers, leading them back to the dinner hole. You see the shuttle car operator and he tells you the section foreman is now on the phone. He also tells you that the superintendent has come in to check things out, but they still have not pinned down the source of the fire. Everyone is now in the dinner hole except the section foreman. What can you do to help at this point?

6. YES NO Tell everyone to walk down to the phone and try to find out what is going on.

7. YES NO Tell everyone to go get on the mantrips immediately.

8. YES NO Go to the SCSR cache on the section, gather the SCSRs, and distribute them.

9. YES NO Ask someone to bring the tether rope from the SCSR cache.

10. YES NO Verify the section power has been cut.

11. YES NO Do a head count.

12. YES NO Go get the section foreman now. Everyone is waiting!

13. YES NO Send someone down to warm up the mantrips, and bring the outby mantrip closer to the dinner hole.

When you have responded to all the statements above, go to the next page.

Question C

Several minutes have passed while the section foreman has been trying to get more information about the fire's location. You begin to smell smoke. You look for the foreman's light and see he has not started back to the dinner hole yet. What should you do now?

14. YES NO   Tell everyone to don their SCSRs now and get on the mantrips.

15. YES NO   Tell one of the shuttle car operators to take one of the mantrips and start out of the section.

16. YES NO   Go get the foreman and tell him that smoke is coming into the section.

17. YES NO   Do nothing until the section foreman comes back to the dinner hole with more information.

When you have responded to all the statements above, go to the next page.

Question D

You tell everyone to don their SCSRs and go to the mantrips. In the meantime, you don your SCSR and go to find the section foreman. You find him at the phone and show him a note pad on which you have written "Smoke Coming." The foreman tells you that he talked to the communications person, but has not been able to find out any more information. What should you do now?

18. YES NO   Stay at the phone to try to get more information while the foreman heads back to the mantrips to stay with the crew.

19. YES NO   Gesture to the foreman to don his SCSR and that you need to start out of the section before the smoke gets any heavier.

20. YES NO   Signal to the foreman that you are going to check the returns and secondary escapeway for smoke.

21. YES NO   Go back to the mantrips and signal the crews that you are waiting to get more information on the fire before leaving the section.

When you have responded to all the statements above, go to the next page.

Question E

The section foreman agrees that you should leave the section. Just before donning his SCSR, he calls outside to advise that the crews are starting out of the section on the mantrips in the primary escapeway. The foreman also tells the communications person he will also stop at the permanent refuge chamber and call out to check on the fire. You travel with the others outby and reach the permanent chamber at No. 50 crosscut where you meet two outby laborers. The foreman goes into the chamber and calls out to check on the fire, but does not get any new information. The smoke is heavier than when you left the section. What should you do now?

22. YES NO   Suggest to the foreman that everyone should get into the permanent chamber and wait for the arrival of the mine rescue team.

23. YES NO   Suggest that the crews move into the secondary escapeway and continue escape on foot because there may be little or no smoke, and they can follow the belt line.

24. YES NO   Stay on the mantrips and try to make it to the main permanent refuge chamber in North Mains or to the outside.

25. YES NO   Suggest that the crews move into the return and continue escape.

When you have responded to all the statements above, go to the next page.

Question F

Although you suggest that the crew take refuge in the permanent refuge chamber, the miners are determined about getting out. In addition, the section foreman feels you should continue outby because the smoke is not heavy and the visibility is about 40 ft. He waves his finger for you to accompany them. You think about it and decide that, rather than split the crew, you will continue outby knowing you can always come back. You and the others travel another 1,200 ft in the mantrip when suddenly you hit heavy, thick smoke. The visibility is less than 10 ft. The mantrip operator slows down and proceeds cautiously but is bouncing off the ribs because of poor visibility. After traveling a short distance, your mantrip rams an abandoned road grader. At this point, miners begin exiting the mantrips and start to scatter. What should you do now?

26. YES NO   Help the foreman gather miners back together. Make sure all miners have their SCSRs on. Then gesture to everyone that it's time to head back inby to the permanent chamber at No. 50 crosscut.

27. YES NO   Find the nearest door and check the secondary escapeway.

28. YES NO   Offering to bring up the rear, signal the boss to lead the crew out the primary escapeway.

29. YES NO   Gather up the extra SCSRs from the mantrips.

30. YES NO   Take the tether rope you brought to use as a linkline between miners.

When you have responded to all the statements above, go to the next page.

Question G

The section foreman checks for smoke in the secondary escapeway and the carbon monoxide (CO) level; the CO level is 45 ppm. He yells back that the smoke is lighter, and the visibility is about 50 ft. Even though it's more than a mile uphill to the main and two miles to the portal, the section foreman puts his SCSR back on and gestures to follow him out the secondary escapeway. Everyone grabs an extra SCSR. Several miners start to go with the foreman. Sam, a mechanic who is overweight and out of shape, takes his SCSR off and says, "I'm already outta breath and we ain't even started to walk. I know I can't make it if I have to walk more than a mile uphill to get out!" What should you do?

31. YES NO   Signal everyone that you are taking Sam and going to the permanent refuge chamber and wait for the mine rescue team to get the fire under control and to come to get you out. Ask who is going with you.

32. YES NO   You don't want to split the crew, so you ask the section foreman to put the question to a vote.

33. YES NO   Gesture to Sam that you will stay back and help him as the crew proceeds out on foot.

34. YES NO   Suggest that someone move the road grader out of the way so you can continue outby on the mantrip.

When you have responded to all the statements above, go to the next page.

Question H

The section foreman, another mechanic, three labor crew workers, and eight miners from the production crew decide to head out the secondary escapeway. One of the remaining miners with you shrugs his shoulders as if saying, "What now?" What should you do?

35. YES NO Signal the miners remaining to get back on the inby mantrip with you and ride back to the permanent chamber.

36. YES NO Indicate to everyone that you are going to walk downhill back to the permanent refuge chamber through the primary escapeway with the smoke at your back.

37. YES NO Signal everyone that you are going to walk back down to the permanent chamber. Motion for them to keep their SCSRs on, grab the tether rope, link together, and follow you into the secondary escapeway where the smoke is lighter.

38. YES NO After having second thoughts, wave to everyone that you have changed your mind and have decided to follow the section foreman.

When you have responded to all the statements above, go to the next page.

Question I

The remaining miners, including the maintenance foreman, choose to go with you to the permanent refuge chamber. With everyone linked to the tether rope, you proceed down the secondary escapeway to the permanent refuge chamber. As you had thought, the smoke is lighter in this entry. When you reach the permanent refuge chamber, you motion for the miners to enter it one person at a time through the 3-ft entrance tube. As the last person to go in, though, you notice that the smoke in the secondary escapeway is becoming heavier. Now you and the crew are in the permanent refuge chamber room. What should you do now?

39. YES NO   Call outside to check on the fire, report who is with you in the permanent chamber, and advise the communications person that the section foreman is coming out the secondary escapeway with the rest of the crew.

40. YES NO   Tell the miners to take off their SCSRs and relax.

41. YES NO   Ask the maintenance foreman to use his multigas detector and periodically check the atmosphere in the chamber for oxygen deficiency, carbon monoxide, and methane.

42. YES NO   Tell miners that, if they are hungry or thirsty they can open the food and water provisions and help themselves.

43. YES NO   Tell all but one miner to turn off their cap lights to conserve the batteries.

When you have responded to all the statements above, go to the next page.

Question J

You call outside to tell the communications person you are in the chamber with the other miners and you give their names. You learn from the communications person that a power center in the North Mains was on fire, but that the fire is under control and should be extinguished shortly. You are also told that it may be three to four hours before the smoke clears and the rescue team reaches your location. You are also told that no one has heard from the 9 Left foreman or other crew members. You then convey this information to the miners with you in the chamber.
Hearing this, Man Mountain, one of the labor crew members, says: "I ain't stayin' in this hole no three or four hours!! If the fire's out, I'm leaving!!" What should you do now?

44. YES NO   Put your SCSR back on, go out of the chamber, and check both the secondary and primary escapeways for smoke. If the smoke is clearing, lead Man Mountain and the rest of the crew out.

45. YES NO   Tell Man Mountain to go ahead and leave, but that you are staying in the chamber.

46. YES NO   Tell Man Mountain that he would have to travel through smoke nearly a mile uphill just to reach the mains.

47. YES NO   Remind Man Mountain that the section foreman and the rest of the crew, who decided to travel out the secondary escapeway, have not been heard from and that he's taking a big risk by leaving.

When you have responded to all the statements above, go to the next page.

Question K

You notice Man Mountain is sweating heavily and is visibly agitated. His buddy, Shorty, is watching Man Mountain's behavior and looks scared. Shorty seems to be short of breath and says he thinks he is having a heart attack. The rest of the crew is very quiet. What should you do?

48. YES NO Ignore them, figuring they will calm down.

49. YES NO Mention to everyone that this is a stressful situation and that some symptoms of stress are expected.

50. YES NO Begin CPR on Shorty.

51. YES NO Tell everyone it is normal to be upset in a situation like this and most people react physically and emotionally like sweating a lot, getting an upset stomach, experiencing an increased heart rate, and feeling scared, angry, or quiet.

52. YES NO Suggest everyone take three deep breaths to increase oxygen in their bodies and decrease the adrenalin.

When you have responded to all the statements above, go to the next page.

Question L

After thinking about what you told him, Man Mountain decides to stay in the permanent refuge chamber with you. Shorty calms down and says he feels better. After about three hours, you hear the outside door of the entrance tube open. In a minute, the inner door opens and a mine rescue team captain enters. He tells you that the fire is out, the smoke has cleared, and that mantrips are waiting to take you and the others outside. Someone asks if the other crew members are out yet. The rescue captain informs you that the others did not fare as well. He says two of the miners were overcome by CO near the mouth of 9 Right and are on their way to the hospital. He also tells you three others got separated from the group and a search for them is underway.

53. Do you feel you made the correct decision by seeking refuge in the permanent refuge chamber? Why or Why not? *Please write your answers below.*

54. When the crews split up and some members decided to escape on foot, what could have been done to persuade the other miners to go to the permanent refuge chamber? *Please write your answers below.*

55. What are the positive aspects of seeking refuge? *Please write your answers below.*

56. What are the negative aspects of seeking refuge? *Please write your answers below.*

When you have responded to all the statements above, go to the next page.

Question M

The three miners who were separated from the group who tried to escape were found dead by the mine rescue teams. Two of the miners were 100 ft from an SCSR cache along the escapeway. The third miner was found behind a barricade he erected out of some brattice cloth. The event is over, but you are having trouble sleeping, and find yourself eating more and drinking more alcohol than usual. You feel depressed and your wife has noticed you are having difficulty concentrating and making decisions, even about the most trivial things. She is worried about you. You have become concerned about your safety and those of your crew when going back underground. What should you do?

57. YES NO   Do nothing; time will heal.

58. YES NO   Be especially careful when driving. Pay attention to the conditions and where you are going.

59. YES NO   Request outside work at the mine for awhile.

60. YES NO   Quit your job.

61. YES NO   If you continue to feel depressed and anxious or are having nightmares, seek counseling from a disaster mental health professional.

End of Problem

Finding Your Score

1.	Count the total number of responses you circled that were marked "YES". Write this number in the box to the right.	
2.	Count the total number of incorrect responses you circled. Subtract this number from 29 and write the difference in the box to the right.	
3.	Add the numbers on lines 1 and 2. Write that number in the box to the right. This is your score on the exercise.	

The best score is 57. The worst score is 0.

www.ingramcontent.com/pod-product-compliance
Lightning Source LLC
Chambersburg PA
CBHW081807170526
45167CB00008B/3357